Impedance matching using a single stub and a transmission line.

- Smith chart approach for simplicity
- Simulation results of example

By: Ain Rehman

Signal Processing Group Inc.

Single stub matching technique:

Refer to Figure 0 below. It shows the technique used to do impedance matching using a single stub connected in shunt with a main line as shown.

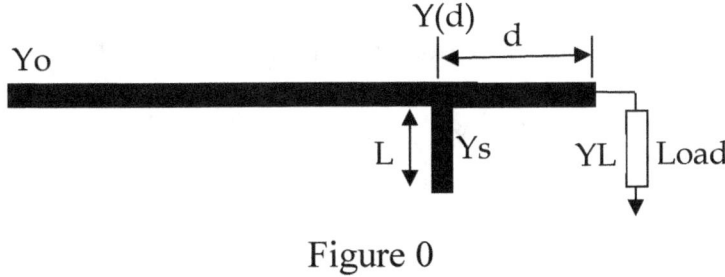

Figure 0

Figure 0 depicts the layout of a single stub matching circuit. The load is YL. The microstrip line has a characteristic admittance of Yo. The load needs to be matched to the characteristic admittance (or the characteristic impedance of the line). A single stub is used in this case that is situated a distance 'd' from the end of the line where the load is connected.

It is well known that the admittance changes as the line is traversed from the load end towards the source. What we want to do is to find a length of line 'd' that will change the load, to where, at the point 'd' it will see the characteristic admittance and a susceptance which can be positive or negative.

In other words at point 'd' we want the admittance to be Yo ± jB. Then we want to connect a stub at 'd' such that its susceptance is opposite in sign to the susceptance generated at 'd'. When this is done the admittance seen at the point 'd' is Yo and the load has been matched to the line. The following is a description of the method to do this using the Smith Chart.

Lets match a load $25 - j50$ using a stub to a 50 Ohm line.

Step 1.0 Put the load admittance on the chart at point DP1.

Figure 1.0

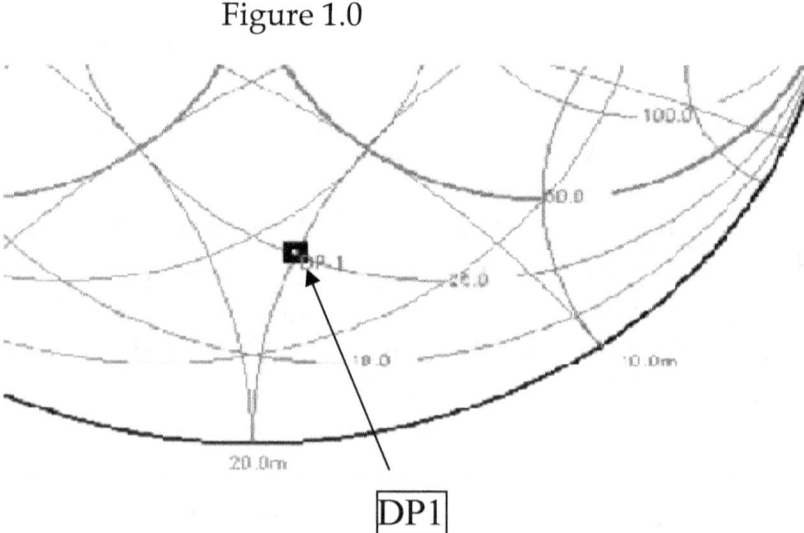

DP1

Note that the chart was set up so as to use admittance and cartesian coordinates. The reason is that shunt stub tuning is being used, so admittance is a more practical choice. So use the immitance chart and set point DP1 as the load as shown.

The details of DP1 are shown below in Figure 2.0

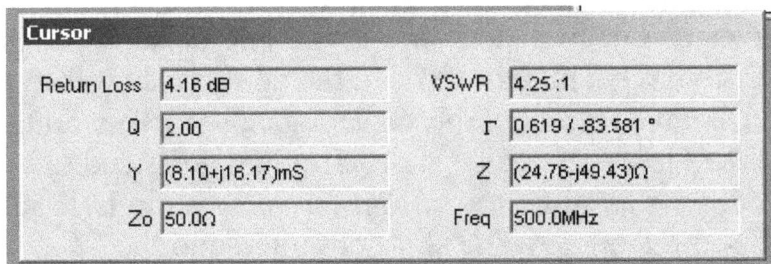

Figure 2.0

Step 2.0   Draw a VSWR circle that passes through DP1 as shown below in Figure 3.0. (Circles are available under 'tools').

Note that the VSWR circle is the circle that shows the variation of impedance around the chart, Note also that this circle intersects the unitary conductance circle (where the admittance is Yo, i.e the reference admittance of 1/50 Siemens. At both intersection points the real part of the impedance becomes the reference admittance while the imaginary part is the susceptance at that point on the line.

Lets examine this a little further. As we move from the load end of the microstrip the admittance starts changing. At a point where the distance from the load end is 'd' the admittance is Yo $\pm jB$. This gives us our first solution of what the distance 'd' is. Now, in order to have a match we need the admittance to be only Yo. So we have to cancel the susceptance term. We do this by using a stub. This stub can be short circuited at its end or open circuited. Lets assume it is a short circuited stub. Lets start at the short circuited end. At the short circuited end the conductance is infinite. This point is at the left extremity of the center diagnol of the chart. So we will start there. Now because we are dealing with only a susceptance term we will move on the periphery of the chart. The data presented below shows what the numbers are: The susceptance at point A is: 31.47mS. We have to cancel this susceptance by making the stub length and type to be -31.47mS.

Figure 3.0

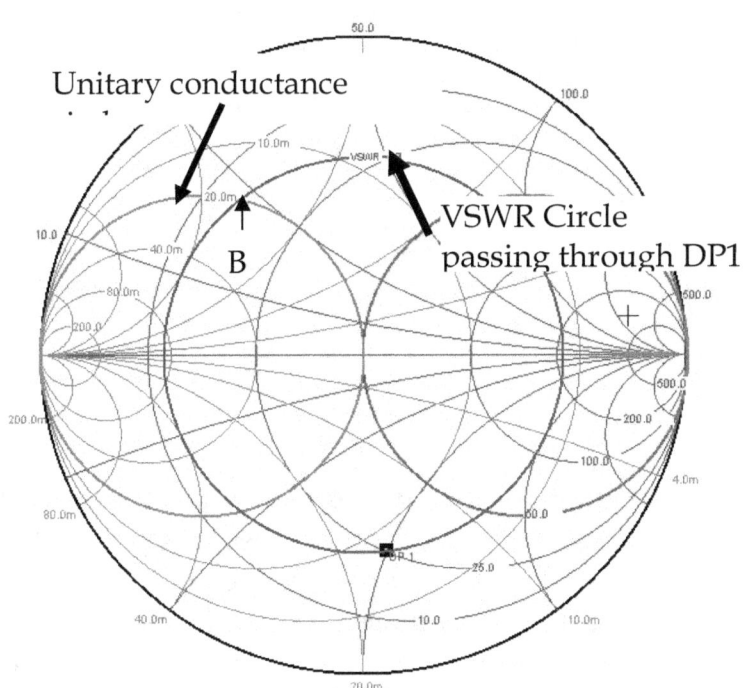

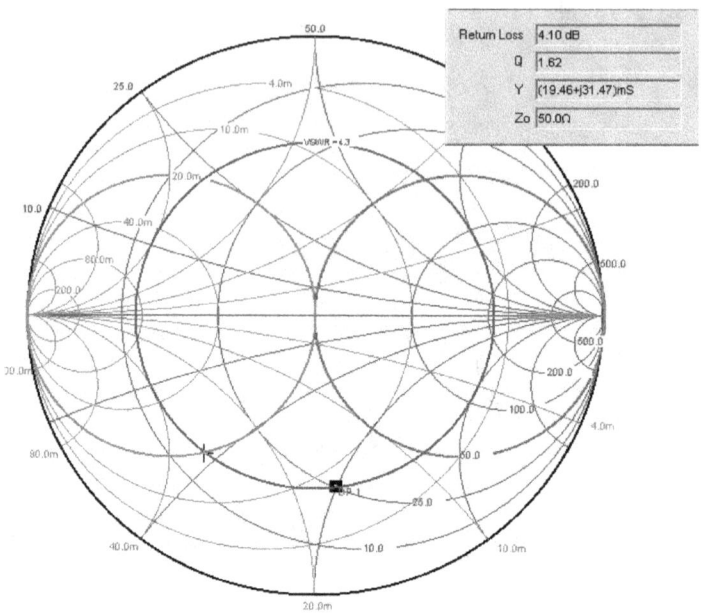

Figure 4.0

A note on the type of stub selected: The analysis and example was based on a shorted stub. You can use an open stub. What kind of stub to use depends on preference and ease of use and tuning and modifying. If you use an open circuit stub then you have to start from conductance of zero (at the right hand side of the chart). Please refer to chart properties in the section on Smith Chart properties.

Angle is 123.94

Stub length is: 180 – 123.94 Deg = 56.06 Deg. Move clockwise around the periphery towards the generator (or source) Verify this using the equation for a shorted stub.

At this point the susceptance is -37.57mS

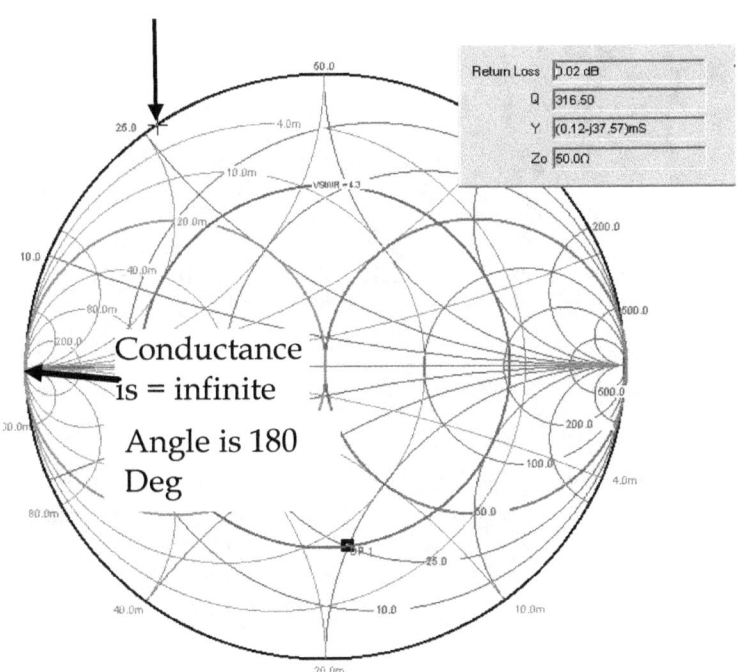

Figure 5.0

The following sections of the book focus on examples of single stub matching and simulations using a public domain simulator available from QUCS. http://qucs.sourceforge.net/

Example 1.0 Match a 75 Ohm transmission line to a load of 24.8+ j7.0 Siemens.

Figure 6.0

75 Ohm TX line

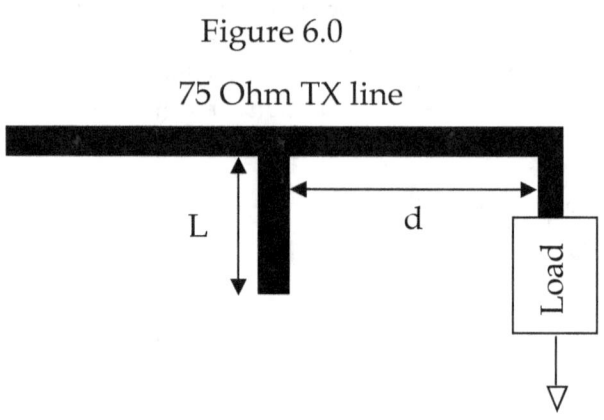

Step 1.0

Place the load YL = (24.8 + j7.0)mS on the Smith Chart as shown below.

Please note:

We are using a Smith Chart tool provided as a demo version by Professor Dellsperger, Juerg Tschirren and Roger Wetzel of Berne Institute of Engineering and Architecture, Switzerland. This tool is available on the web.

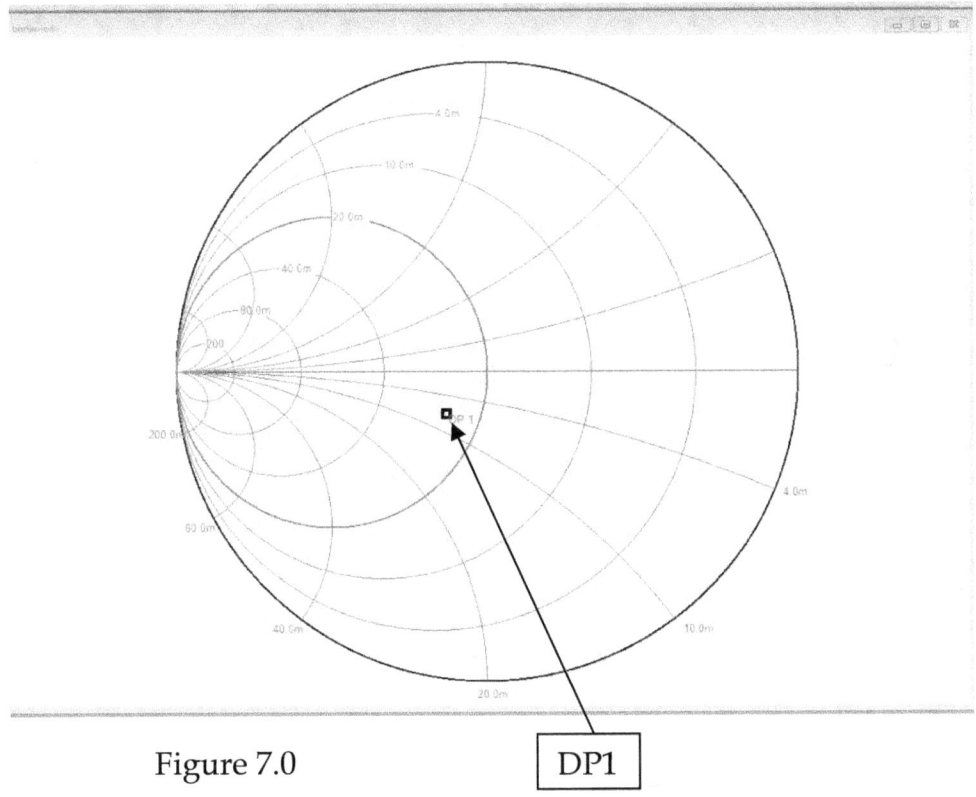

Figure 7.0   DP1

Note that this point is a normalized value of the load admittance. The reference impedance is 50 Ohms. (Reference admittance = 1/50 = 0.02 S or 20 mS.

Step 2:

Draw a VSWR circle passing through DP1. Circles are available under "tools" in the Smith Chart program .

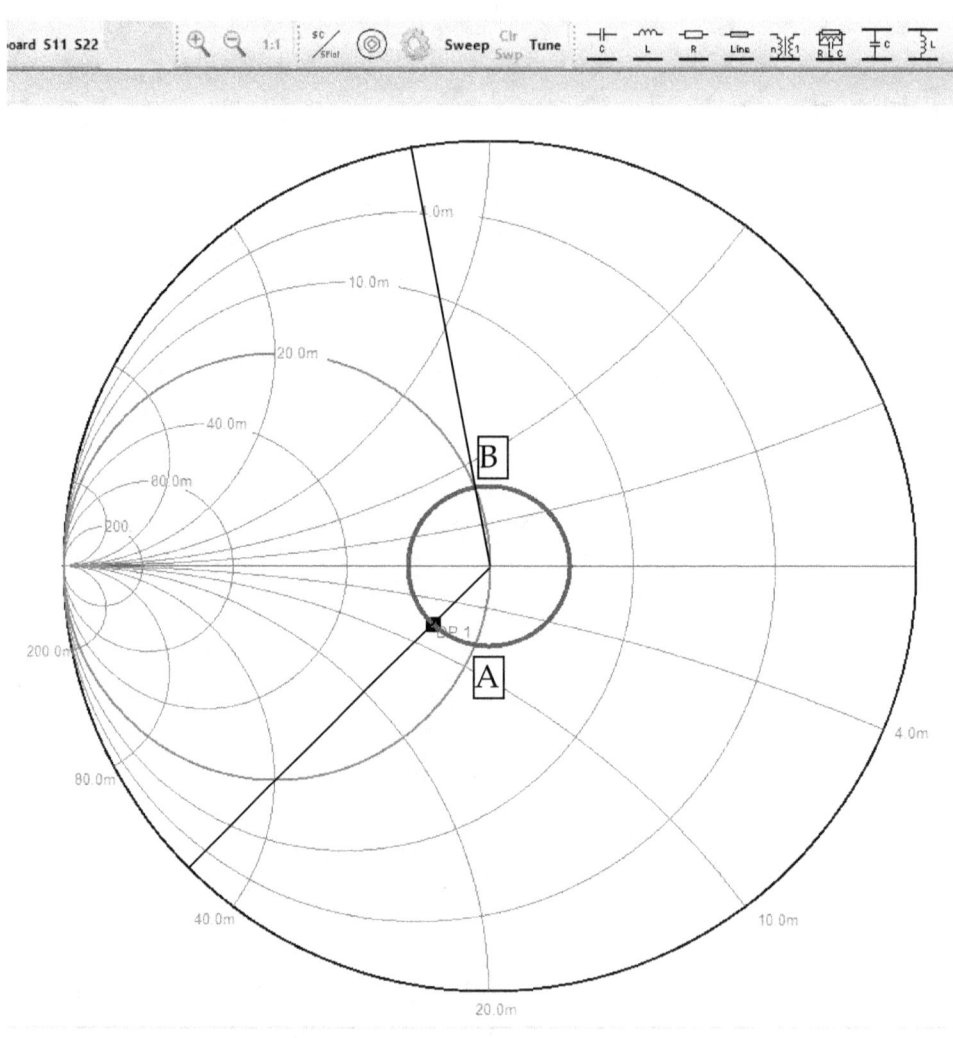

Figure 8.0

Note that the VSWR circle intersects the unitary admittance circle at two points: Point A = (20.0+j7.56)mS and B =(20.0 + j7.46)mS

We choose intersection point B. The change in angle of the reflection coefficient from DP1 to point B is 147 Degrees. Since once around the chart (360 Degrees) is $\lambda/2$ we know what the length of the line 'd' is if we know what λ is. In our case we chose the frequency to be 500 Mhz. (We can also express the length in λ units. In this case it is 0.4 λ). Then we can calculate the wavelength if we know what the relative permittivity of the microstrip substrate is. Let us assume that it is a FR-4 board with a permittivity of 4.5. We can use any calculator to get the wavelength of the signal on this board. Using the calculator from the Signal Processing Group Inc. website we get the wavelength to be: 283 mm. Then the length 'd' is simply : 115 mm.

Once we move to point B we get the admittance to be Yo (the admittance of the line) and a susceptance of j7.46 mS.

We need to cancel this susceptance using a stub connected at point 'd' that will have a susceptance of -j7.46 mS. We can use an open circuited or a shorted stub depending on preference and convenience.

For the sake of this example we will use an short circuited stub. Since we are using a short circuited stub we need to start from the left hand end of the main diagnol of the admittance chart. Here the conductance is infinite, i.e a short.

We need a susceptance of –j7.46 mS. We again use the Smith Chart to do this. As is well known the periphery of the chart shows susceptance (or reactances as the case may be). So we start at the short circuit end and move towards the source.

With reference to Figure 9.0, we note that the susceptance at the point K is –j7.46 mS. The length of the stub is given by the arc of movement on the periphery of the chart to get to –j7.46 mS. This length is:

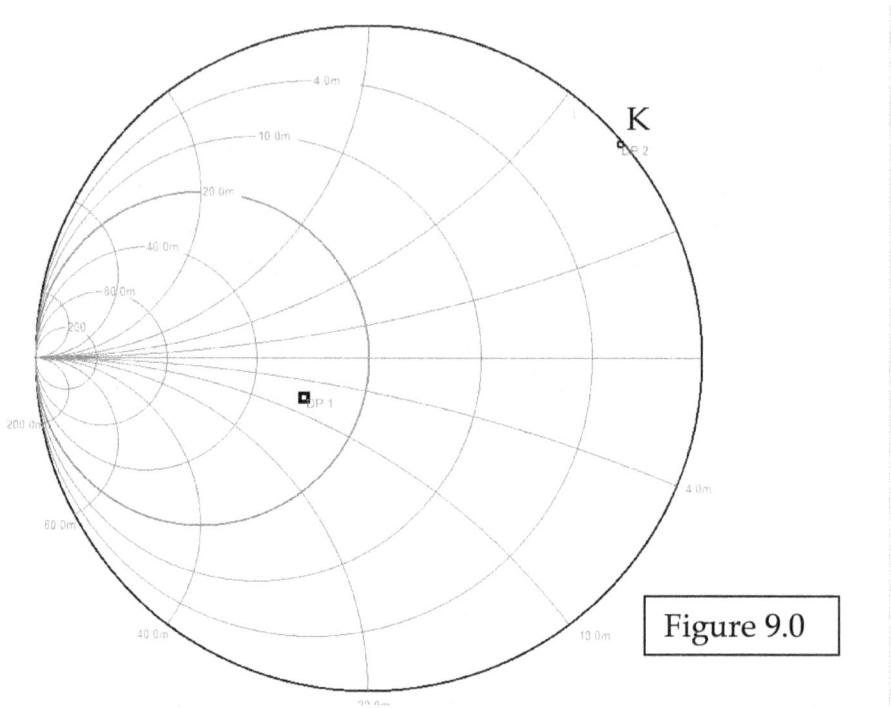

Figure 9.0

If we calculate the length of the stub, we see it is: 0.66λ (a change in angle of 220 degrees from the open end of the stub to its length. The reason is that the angle at the open end is 0 degrees. Then as we move in a clockwise direction from there we first move 180 degrees then another 40 degrees to get to the required susceptance of –j7.46 mS. This, then, gives us the final number we are looking for. So we have the distance 'd' and the length of an open circuited stub. It appears from this analysis that a short circuited stub would be shorter in length. That is left as an exercise for the reader.

The next section of this book shows simulation results using these initial numbers and any small modifications needed to meet the specifications.

Please refer to Figure 10.0. This is the reflection coefficient simulation for the single stub circuit design above.

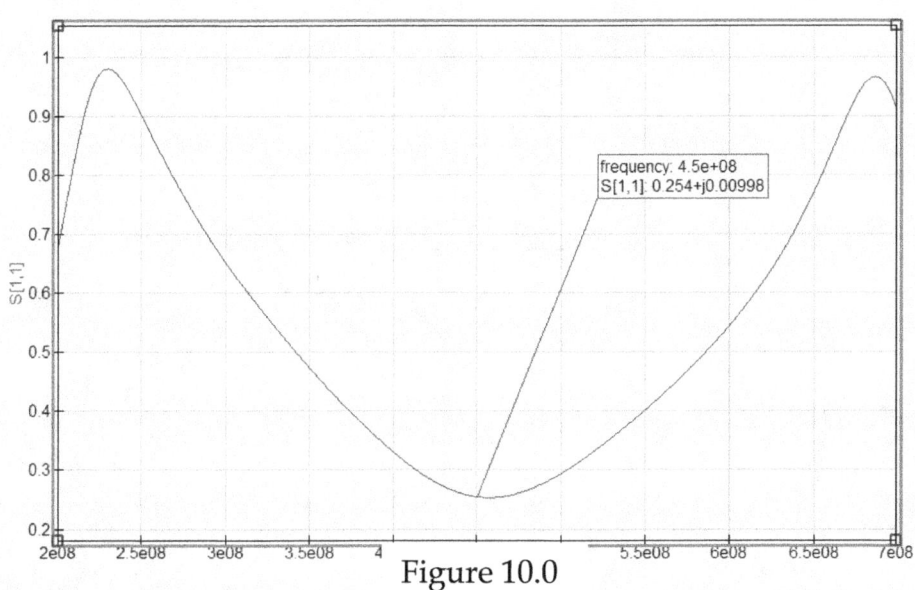

Figure 10.0

Note:
1.0   The reflection coefficient is 0.2. Which is a VSWR of 1.5.
2.0   The frequency of match is not exactly 500 Mhz because this
       was a first cut simulation without any tweaks.
3.0   With a slight amount of tweaking the frequency of match
       can be shifted to 500 Mhz.

The schematic that was used is shown below in Figure 11.0.

The substrate used for the circuit was a FR-4 PCB, with a thickness
of 1.7mm, with 0.0347 mm thickness of 1 oz copper on both sides.

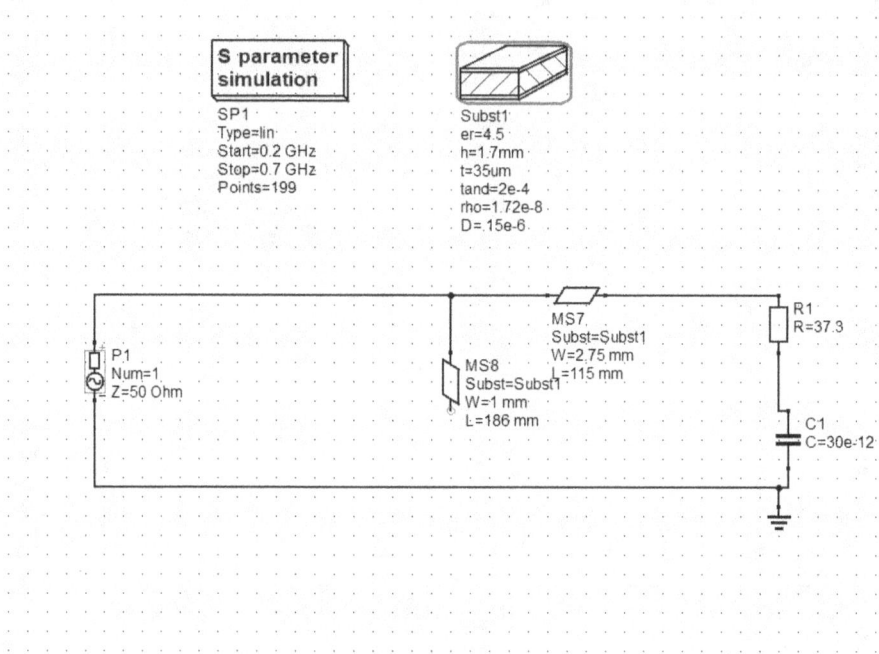

Figure 11.0

The netlist of the circuit used for simulation is shown below:

Qucs 0.0.19 C:/single stub matching book images/qucs_sim.sch

R:R1 _net0 _net1 R="37.3" Temp="26.85" Tc1="0.0" Tc2="0.0" Tnom="26.85"

Pac:P1 _net2 gnd Num="1" Z="50 Ohm" P="0 dBm" f="0.5 GHz" Temp="26.85"

C:C1 gnd _net0 C="30e-12" V=""

SUBST:Subst1 er="4.5" h="1.7mm" t="35um" tand="2e-4" rho="1.72e-8" D=".15e-6"

MLIN:MS7 _net2 _net1 Subst="Subst1" W="2.75 mm" L="115 mm"

Model="Hammerstad" DispModel="Kirschning" Temp="26.85"

MLIN:MS8 _net3 _net2 Subst="Subst1" W="1 mm" L="186 mm"

Model="Hammerstad" DispModel="Kirschning" Temp="26.85"

.SP:SP1 Type="lin" Start="0.3 GHz" Stop="0.7 GHz" Points="199"

Noise="no" NoiseIP="1" NoiseOP="2" saveCVs="no" saveAll="no"

Some transmission line facts are listed below:

Transmission line facts:

0.0 Transmission lines that have electrical lengths *less than 90 Degrees* behave *inductively for short circuited loads* and *capacitively for open circuited loads.*
1.0 A *RF short circuit* can be produced *at any point* in a circuit by using a *short circuited transmission line with a half wavelength electrical length.* This short repeats itself every multiple of a half wavelengh.
2.0 A transmission line's input impedance is always equal to the termination at adjacent ends of the line if the characteristic impedance of the line is *the same as* the termination.

3.0 Transmission lines have *large transformation capabilities*. A short circuit can be transformed to an open circuit by using a *90 Degree* long line. The same is true of an open circuit. An open circuit can be transformed into a short circuit by a *90 Degree* line.

4.0 Cascaded transmission lines form *concentric circles* on a normalized Smith Chart if the load impedance is normalized to the characteristic impedance of the transmission line.

5.0 Parallel open and short circuited stubs behave *inductively or capacitively* as long as their electrical length is *less than 90 Degrees.*

6.0 Parallel stubs *always* move on the constant conductance circles on the Smith Chart.

7.0 A correctly selected combination of a *parallel stub and cascade transmission line* can be used to *transform any point* on the Smith Chart to any other point. The topology of this combination is dependent on the relationship of the two points. Sometimes the cascade line can be used first followed by the stub, while at other times the stub is followed by the cascaded line.

Open and shorted microstrip line characteristics: Let us examine how open and shorted microstrip lines behave based on their lengths as a function of effective wavelength of the signal.

Open line:
When a line is open ended then at all *odd* quarter – wave points (1/4, 3/4 etc) the impedance is *minimum*. The line acts like a series resonant circuit. So if one wishes to establish a RF short circuit to ground (for example) at some point on a microstrip line, then an open ended, quarter-wave (or odd multiples) of an open ended line connected to that point will drive the line to very low impedance and short the frequency that is used to ground. See Figure M.5 below.

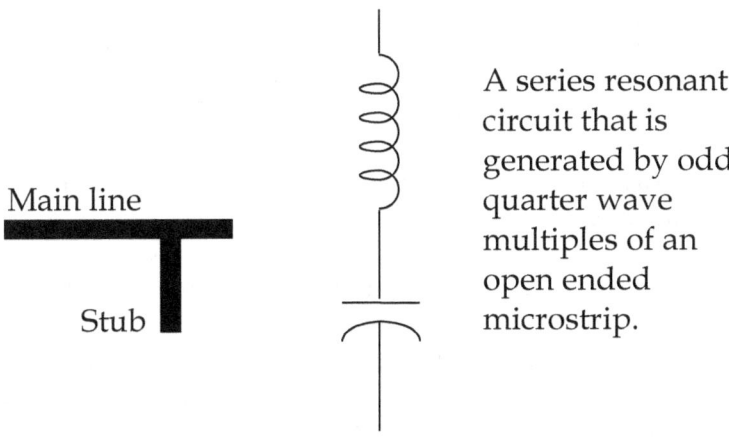

A series resonant circuit that is generated by odd quarter wave multiples of an open ended microstrip.

Figure M.5

At all *even* quarter – wave points (1/2, 1, 3/2 etc) the impedance is a *maximum.* The line behaves as a parallel resonant circuit. The line at resonance has a very high impedance. Please refer to Figure M.6 below.

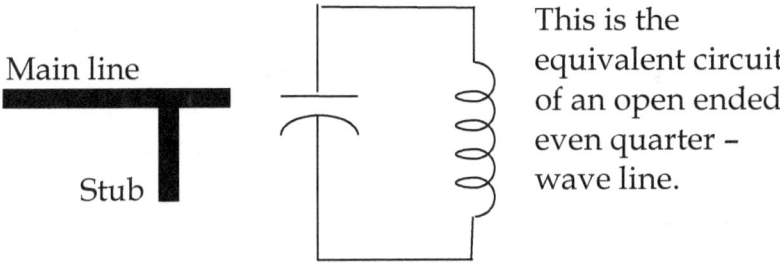

This is the equivalent circuit of an open ended even quarter – wave line.

Figure M6.0

An earlier discussion in this book addressed open and shorted stubs that are less than a quarter wavelength long. An open ended line that is less than a quarter – wave long acts as a capacitance.

A open ended line that is between a 1/4 to a 1/2 wavelength long acts as an inductance. An open ended line that is 1/2 to 3/4 wavelength long acts as a capacitance etc.

Shorted line:

When the line is shorted at one end (RF short) then at *odd* quarter – wavelength points, it acts like a parallel resonant tuned circuit. Thus at the frequency of resonance the impedance is very high.

At *even* quarter - wave points the line acts as a series resonant circuit and generates a low impedance at the resonant frequency

Resonant shorted lines also can act as pure capacitances and inductances. A shorted line that is less than a 1/4 wavelength long acts as an inductance. A shorted line that is between a 1/4 to a 1/2 wavelength long acts as a capacitance. From 1/2 to 3/4 wavelength it acts as an inductance and from 3/4 to 1 wavelength long it acts as a capacitance and so on.

It should be clear that appropriately chosen line lengths can be used as parallel – resonant or series – resonant , inductive or capacitive circuits. This is a very powerful conclusion and is very useful for the design engineer.

Open circuit microstrip lines and terminations: The following are relevant facts:

O1.0   The voltage at the open end is maximum but the current is minimum.

O2.0   The distance between two adjacent zero current points is 1/2 wavelength.

O3.0   Distance between two alternate zero current points is 1 wavelength.

O4.0   At any frequency the voltage on the line is minimum or zero at 1/4 wavelength from the end of the line.

O5.0   Voltage peaks occur at the end of the line, at 1/2 wavelength from the end and every 1/2 wavelength after that.

Short circuit microstrip lines and terminations: The following are relevant facts:

S1.0   The voltage at the end (terminated end) is obviously zero.

S2.0   The voltage is maximum at 1/4 wavelength from the end and alternately maximum at every 1/4 wavelength after that.

Microstrip lines terminated in Zo: The following are relevant facts:

Z1.0   In a properly terminated line the voltage and curent will be constant along the line unless there are some losses such as resistive or otherwise.

Z2.0   If there are losses in the line (and there are sure to be some) the voltage and current will become smaller along the line towards the termination.

A point worth noting is that if a line is designed to be a 1/4 wave at frequency fo then it would change its characteristics at a frequency 2fo and 1/2fo etc. This fact must be borne in mind. It could be an advantage or a disadvantage depending on the conditions.

Cross – junction stubs: This is a technique that can be used advantageously in the use of stubs. *When the impedance of the stub is really low, i.e it has a large width then one solution is use two stubs in parallel connected on both sides of the main line.* The impedance of each of the equivalent stubs is twice the impedance of the original stub. See figure M.7.0 below.

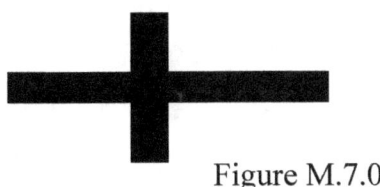

Figure M.7.0

References:

1.0 Practical Impedance Matching, by Ain Rehman, published by Signal Processing Group Inc.
2.0 QUCS public domain simulator

3.0 Smith Chart program, demo version developed by :

V 1.91
This program has been developed by Prof. Fritz Dellsperger,
Juerg Tschirren and Roger Wetzel
© 1995 - 2000 by Berne Institute of Engineering and Architecture

4.0 Various calculators developed by Signal Processing Group Inc. for RFMW and analog design available from www.signalpro.biz, the Signal Processing Group Inc. website.

Notes:

Notes:

www.ingramcontent.com/pod-product-compliance
Lightning Source LLC
Chambersburg PA
CBHW071205220526
45468CB00003B/1165